Garden Plants
Valuable to Bees

Jointly published by
International Bee Research Association

INTERNATIONAL BEE
RESEARCH ASSOCIATION

and by Northern Bee Books

Northern Bee Books

Garden Plants Valuable to Bees: Edited by Eva Crane

Principle collaborators
Mary F Mountain, MSc, DipLa, Rosemary Day, BSc (IBRA) Christine Quartley, PhD, (BSBI), Alison Goatcher (RHS)

Copyright: 2017 by International Bee Research Association

All rights reserved. No part of this publication may be reproduced, stored in a retrieval system, transmitted in any form or by any means electronic, mechanical, including photocopying, recording or otherwise without prior written consent of the copyright holders.

First published in 1981
Reprinted in 1992, 2003, 2008
This is a reprint of the 2008 edition jointly published by
© IBRA
www.ibrabee.org.uk
and
Northern Bee Books
www.northernbeebooks.co.uk

ISBN: 978-0-86098-287-6

New cover design and artwork by DM Design & Print

Front cover (honeybees foraging on Limnanthes flowers) photo by *David Heaf*

Printed by Lightning Source UK

Garden Plants
Valuable to Bees

Jointly published by
International Bee Research Association

INTERNATIONAL BEE
RESEARCH ASSOCIATION

and by Northern Bee Books

Northern Bee Books

Garden Plants Valuable to Bees

CONTENTS

Preface   3

Information about the tables   5

TABLES

    Herbaceous plants   7

    Plants for the rock garden   15

    Culinary and aromatic herbs   17

    Bulbs, corms and tubers   19

    Annual and biennial plants   22

    Plants for the wild garden   27

    Climbers   34

    Trees   36

    Shrubs   42

Reading list   55

PREFACE

There are books about garden plants for birds, and for butterflies, and here at last is one on garden plants for bees. More than any other member of the animal kingdom, bees can enrich our gardens. Not only does their contented hum have a pleasant sound; watching their foraging behaviour is a source of endless pleasure to bee lovers. If we can attract bees into our ornamental gardens, they will pollinate the fruit in our kitchen gardens, and we shall get larger crops and more perfectly formed fruits.

If we put a hive or two of bees into our own gardens we shall get a honey crop too. Bees often fly several kilometres to collect the nectar that they make into honey, which comes from plants in an area of several hundred hectares or acres. The honey is derived not so much from the garden plants as from more massive flowerings – of a lime tree avenue, a long hawthorn hedge, or a wilderness full of brambles and willowherb. Pollen provides protein needed for rearing the next generation of bees.

The importance of garden plants yielding nectar and pollen is that together they provide a continuous food supply, form willows and crocuses in early spring to ivy in late autumn. Colonies of bees need food all through the active season, so that they can develop and rear new bees that will collect the harvests when they come – perhaps during a week or two in May, and rather longer in July. This continuous food supply used to be provided by pastures that came into flower before they were cut, by verges and hedgerows, and by abundant weeds. Nowadays the efficiency of agriculture has greatly reduced these resources for the bees. But gardens designed for ease of management are most valuable in providing day-to-day food supplies that the countryside no longer yields. This provision is especially valuable in what beekeepers refer to as the June gap, when there is often a lull in the intensity of flowering.

The garden plants selected for inclusion in this book are not only valuable to bees: they are worth growing in their own right, and have been chosen for their horticultural merit.

The booklet "Trees and shrubs valuable to bees" by M F Mountain was published in 1965 and again in 1975. Its wide use led to many requests for a similar publication on other garden plants, which this book now provides. Since "Trees and shrubs" is again out of print, the material from it has been incorporated into the final three sections of the present book: climbers, trees and shrubs. Entries in the first six sections, which comprise the major contents, were obtained as follows. They were initially drafted by Miss Mountain, then sent to nine experienced beekeeper/gardeners in different parts of England, who recorded their own observations during two seasons (1979, 1980).

At IBRA the total information was co-ordinated by Rosemary Day and Judith Dolby prepared the camera-ready copy. Botanical names were validated by Dr. Christine Quartley on behalf of the Botanical Society of the British Isles and cultural details were supplied by Alison Goatcher at the Royal Horticultural Society.

The nine beekeeper/gardeners who worked as observers in 1979 and 1980 were: Mildred D. Bindley and David Charles, Somerset; Arthur G. Eames, Derbyshire; Dr. I. Keith Ferguson (Kew), Surrey; R.O.M. Page, Dorset; R.M. Payne, Avon; Rev. Patrick Rowley, South Yorkshire; Betty M. Showler, Bucks.; Ken Stevens, Devon. We are most grateful to all of them, as well as to the principal collaborators listed in the title page, for their work.

Eva Crane, Director
International Bee Research Association

First edition 1981
Reprinted and updated 2008

INFORMATION ABOUT THE TABLES

The lists exclude most crop plants, most fruits and all very invasive plants that are difficult to control. Where single and double flowered cultivars of a plant are available, single flowered cultivars are indicated, since these are more useful to bees.

The lists are designed for Britain and Ireland and also for neighbouring parts of continental Europe: especially Denmark, Belgium, France, Germany and the Netherlands. The flowering season listed is that for Southern England in a "normal" year. "Ordinary garden soil" is neither too acid nor too alkaline (ph 5-7) and contains some humus-forming material.

Under "Notes" the order of entries is usually: description of plant (form of plant/ leaves; arrangement, individual shape, colour, perfume of flowers); cultural requirements (soil type, moisture/drainage; aspect, sun/shade); other points of interest. The following abbreviations are used:

cv(s)	cultivar (cultivated variety)
esp	especially
lv(s)	leaf, leaves
ev	evergreen
posn(s)	position(s)
fl(s)	flower(s), flowering
spp	species
fr(s)	fruit(s)
/	or

The <u>height</u> of plants is given in metres (m); 1m = 100cm.

The final column shows the value of the plant to the bees (meaning honeybees, hive bees); N and P indicate that they collect nectar and pollen, respectively. Plants especially used by bumble bees are marked B. The absence of all three entries (N, P, B) for a few plants reflects insufficiency of observations as to what it is that bees collect from the plant. Garden plants are not normally a source of honeydew and no records are included here of its collection by bees.

The <u>wild garden</u> is regarded as any area planned to provide – with minimal maintenance – a succession of flowers and other sources of interest and attraction, that are not suitable for the cultivated garden. Some of the plants listed for the wild garden are useful to birds or butterflies as well as to bees and a few are escapes and volunteer crop plants. Most entries here would be regarded as weeds outside the wild garden.

Climbers, trees and shrubs are deciduous unless designated evergreen. Climbers are not free standing and require support. All other plants with a trunk are grouped under trees. It is unwise to plant trees listed as growing to height of 30 ft (10 m) or more, unless the garden is a spacious one. If the spread of a tree or shrub differs greatly from its height, this is indicated in the notes.

HERBACEOUS PLANTS

Botanical name	Common name	Flowers	Notes	Height Metres	Value
Acanthus	bear's breeches	Jul-Aug	spectacular form with bracts and fls; sun/light shade	0.9-1.2	B
Achillea	yarrow	Jul-Aug	many spp and cvs; well drained	0.1-1.2	N B
Anchusa azurea	large blue alkanet	Jun-Aug	several cvs, fls bright blue; fertile well-drained soil, sun	0.9-1.5	N B
Anthemis tinctoria	ox-eye chamomile	Jul-Aug	many hybrids with A. sancti-johannis; well-drained soil, sun	0.8	N B
Armeria	thrift	May-Aug	pink/red/white fls; well drained soil, full sun	0.05-0.6	N P B
Aster			daisy-like fls; fertile soil, sun		N P
A. amellus	Michaelmas daisy	Aug-Sep	several cvs, lavender/pink/rose fls	0.5-0.6	N P
A. novae-angliae	Michaelmas daisy	Sep-Oct	several cvs, pink/lavender fls	1.2-1.5	N P
A. novi-belgii	Michaelmas daisy	Sep-Oct	many cvs, wide range of shades	0.23-1.2	N P
A. sedifolius = A. acris	Michaelmas daisy	Aug-Sep	lavender blue fls	0.6-0.8	N P

7

Garden Plants Valuable to Bees

Botanical name	Common name	Flowers	Notes	Height Metres	Value
Caltha palustris	marsh marigold	Feb-Jan	profuse cup-shaped yellow golden fls; water up to 15cm deep/loamy moist soil	0.3-0.5	N P B
Campanula lactiflora	milky bellflower	Jun-Jul	several cvs, lavender-blue bell shaped fls; fertile well-drained soil	0.9-1.5	N P
C. persicifolia	peach-leaved	Jul	white/blue/purple fls; well-drained fertile soil.	0.3-0.9	N P
Centaurea Montana	perennial cornflower	Jun	blue fls, other cvs with white/pink fls; fertile well-drained soil, sun	0.5-0.6	N B
Chrysanthemum	Korean Chrysanthemum	Oct	many shades; fertile well-drained soil, some lime, sun	<0.9	P
C. coccineum (single cvs)	pyrethrum	May-Jun	wide range of shades; well-drained soil, sun	<0.9	P
C. maximum = Leucanthemum maximum	Shasta daisy	Jun-Aug	single white fls; fertile well-drained soil preferably with lime, sun	0.8-0.9	B
Cichorium intybus	chicory	Jul-Oct	sky-blue dandelion-like fls, also pinkcv; fertile soil with some lime, hearted head of lvs used in salads	0.9	N P B

8

Botanical name	Common name	Flowers	Notes	Height Metres	Value
Coreopsis grandiflora	coreopsis	Jul-Aug	several cvs, bright yellow fls; well-drained soil, stake plants early on	0.3-0.5	N P
C. verticillata		Jun-Sep	fern-like lvs, yellow fls; well-drained soil does not need staking	0.5-0.6	N P
Cynara scolymus	globe artichoke	Jun-Sep	very decorative, edible fl heads; well-drained fertile soil, shelter, sun	<1.5	B
Dictamnus albus	burning bush	Jun-Jul	aromatic plant, white/pink/red fls; well-drained soil preferably with lime, sun	0.6	N P
Doronicum plantagineum	leopard's-bane	Mar-Apr	golden-yellow fls; deep moist soil; sun/pt shade	0.6	N P B
Echinops ritro	globe thistle	Aug	steel-blue fls to dry for indoor decoration well-drained soil, sun	0.9-1.2	N B
Erigeron	fleabane	Jun-Aug	several spp and cvs, daisy-like fls in many shades; moist well-drained soil, sun	0.08-0.6	N B
Eryngium maritimum	sea holly	Jul-Sep	cone-shaped metallic blue fls suitable for drying; well-drained soil, sun	0.3-0.5	N B
E. tripartitum		Jul-Sep	grey-blue globular fls; well-drained soil, sun, needs staking	0.8	N B

Garden Plants Valuable to Bees

Botanical name	Common name	Flowers	Notes	Height Metres	Value
Eupatorium		Jul-Sep	white/red/purple/pink fls; moist soil, sun/pt shade	0.6-1.8	N P
Gaillardia hybrids	blanket flower	Jun-Oct	yellow and red daisy-like fls; light well-drained soil, sun	0.6-0.8	N P
Geranium ibericum		Jun-Aug	violet-blue fls; well-drained soil, sun/pt shade	0.5-0.6	N
G.phaeum	dusky cranesbill, Mourning widow	Jun-Jul	dark maroon fls, cvs with white fls; tolerates deep shade	0.6	N
G. pratense	meadow cranesbill	Jul-Sep	blue/violet-blue fls, other cvs white/light blue; well-drained soil, sun/pt shade	0.5-0.6	N
Geum	avens	May-Jun	many spp. long lasting bright coloured fls; ordinary garden soil, sun/pt shade	0.15-0.6	P
Gypsophila paniculata (single cvs)	baby's breath	Jul-Aug	numerous tiny white fls; well-drained soil, sun	0.9	N
Helenium	sneezeweed	Jul-Aug	many hybrids and cvs, yellow/red/orange fls; ordinary soil, sun	1.2-1.8	N P B
Helianthus	sunflower	Jul-Oct	yellow fls; well-drained soil, sun	0.9-3.0	N P

10

Botanical name	Common name	Flowers	Notes	Height Metres	Value
Helleborus	Christmas rose, Lenten rose, hellebore	Dec-Mar	white/green/purple fls; well-drained soil several spp like shade	0.3-0.6	N
Heuchera	coral flower	Jun-Sep	many hybrids and cvs, tiny red/pink/white fls; light well-drained soil, sun/pt shade	0.3-0.5	N
Impatiens noli-tangere	touch-me-not	Aug-Sep	yellow fls spotted with red inside; moist soil	0.5	N P B
Iris unguicularis = I. stylosa	Algerian iris	Oct-Mar	fls soft lavender with a yellow blaze on falls; well-drained soil, sun	0.23	N
Liatris spicata	blazing star,	Sep	pink-purple fls, tuberous roots; ordinary soil, sun	0.6-0.9	N
Ligularia dentate = Senecio clivorum		Jul-Aug	clusters of orange-yellow fls; ordinary soil, preferably moist	0.9-1.5	N P
Limonium	sea lavender, statice	Jul-Sep	lavender blue/pink/purple fls suitable for drying; well-drained soil, sun	0.3-0.6	N
Lobelia cardinalis	cardinal lobelia,	Jul-Aug	scarlet fls; rich moist soil, pt shade, shelter, winter, protection	0.8	B
Lupinus	lupin	May-Jul	many small fls in colourful spire-like racemes; neutral acid soil, sun/pt shade	0.9-1.5	P B
Lysimachia vulgaris	yellow loosestrife	Jul-Sep	yellow fls; moist soil, shade	0.6-0.9	P B

11

Garden Plants Valuable to Bees

Botanical name	Common name	Flowers	Notes	Height Metres	Value
Monarda didyma	sweet bergamot bee balm, Oswego tea	Jun-Sep	scarlet/purple/pink/white fls; aromatic lvs can be used in tea; moist soil, sun/pt shade	0.6-0.9	B
Nepeta x faassenii	catmint	Jun-Aug	spikes of lavender-blue fls; well-drained soil, sun/pt shade	0.3-0.5	N
Oenothera	evening primrose	Jun-Sep	slightly scented yellow fls; well-drained soil, sun	0.3-1.2	P
Paeonia (single cvs)	paeony	May-Jun	handsome white/yellow/pink/red fls; moist but well-drained soil, sun/pt shade, resents root disturbance	0.5-1.8	P B
Papaver orientale	oriental poppy	May-Jun	large scarlet fls, cvs with pink/white fls; well-drained soil, sun	0.6-0.9	P B
Penstemon		Jun-Jul	many hybrids, scarlet/rose/purple fls; well-drained soil, full sun	0.3-0.6	N B
Polemonium caeruleum	Jacob's ladder, Greek valerian	Apr-Jul	blue/white fls; rich soil, sun/pt shade	0.3-0.6	N P
Polygonatum	Solomon's seal David's harp	May-Jun	attractive form, white fls; almost any type of soil, sunny posn	0.6-1.2	N B
Polygonum affine		Jun-Sep	red fls; ordinary soil, sun, can be very invasive	0.15-0.23	N P

Botanical name	Common name	Flowers	Notes	Height Metres	Value
P. amplexicaule		Jun on	free-flowering, red fls; rich moist soil, sun/pt shade	0.9-1.2	N P
P. campanulatum		Jun-Sep	shell-pink fls; rich moist soil, light shade, cut fl stems to ground level in Oct	0.8-1.1	N
Primula vulgaris elatior	polyanthus	Apr-May	many different colours; fertile moist garden soil, sun/pt shade	0.13-0.23	B
Pulmonaria saccharata	soldiers & sailors lungwort	Mar-Apr	fls open pink and change to sky-blue; damp soil, shade	0.3	B
Rudbeckia	coneflower	Jul on	fls orange/yellow/rust; well-drained soil, sun	0.3-2.1	N B
R. laciniata	coneflower	Aug-Sep	yellow fls, green central cone; well-drained soil, sun	1.8-2.1	N
Salvia x superba =S.virgata nemorosa		Jun-Sep	purple-blue fls; well-drained soil, sun	0.6	N
Scabiosa	scabious	Jun-Sep	lavender-blue fls, several cvs; fertile well-drained soil, sun	0.5-0.6	N
Sedum spectabile		Sep-Oct	rose/pink fls; well-drained soil, sun	0.3-0.5	N

Garden Plants Valuable to Bees

Botanical name	Common name	Flowers	Notes	Height Metres	Value
Sidalcea	sidalcea	Jun-Sep	several cvs, spikes of pink fls; ordinary soil, sun	0.8-1.8	N
Solidago	golden-rod	Jul-Oct	many spp and cvs, yellow fls; any soil type, sun/pt shade	0.15-1.8	N P
Stachys	woundwort	Jul-Sep	purple fls; well-drained soil, sun/pt shade	0.15-0.3	N B
S. lanata =S. byzantina	lamb's ear, lamb's tongue	Jul	silver lvs, purple fls; well-drained soil, sun/pt shade, may not survive in cold wet seasons, ground cover	0.3-0.5	N
Thalictrum	meadow-rue	Apr-Aug	many spp, attractive lvs, small fluffy fls; rich moist soil, sun/light shade	0.3-1.5	P B
Verbascum	mullein	Jun-Sep	long spikes of yellow fls; ordinary soil, full sun	0.9	N P B
Veronica spicata	spiked speedwell	Jun-Aug	racemes of blue fls, several cvs; well-drained but not dry soil	0.15-0.5	N P

PLANTS FOR THE ROCK GARDEN

Botanical name	Common name	Flowers	Notes	Height Metres	Value
Alyssum saxatile	golden alyssum	Apr-May	golden-yellow fls; well-drained soil, full sun	0.23	N P
Arabis caucasica	white arabis	Feb-Jun	white fls; well-drained soil, pt shade, very invasive	0.23	N P
Aster alpinus		Jul	purple-blue daisy-like fls, cvs in other shades; fertile soil, sun	0.15	N P
Aubretia deltoidea	aubrieta	Mar-Apr	purple fls, several cvs; well-drained soil, likes lime, sun	0.08-0.10	N P B
Campanula carpatica	Carpathian harebell	Jun	blue/purple/white cup-shaped fls; fertile well-drained soil, sun/pt shade	0.23-0.3	N P B
Carpobrotus edulis = **Mesembryanthemum edule**	Hottentot fig	Jun-Jul	fleshy lvs, showy mauve fls; well-drained soil, good by coast, stands salt spray	0.08	N P
Geranium sanguineum lancastriense	bloody cranesbill	Jun-Sep	pink fls with dark veins, other cvs; ordinary well-drained soil, sun/pt shade	0.08-0.10	N P
Gypsophila repens		Jun-Aug	greyish lvs, small white/pink fls; well-drained soil with some lime, sun	0.15	N

15

Garden Plants Valuable to Bees

Botanical name	Common name	Flowers	Notes	Height Metres	Value
Helianthemum nummularium	rock-rose	Jun–Jul	yellow/orange/red/pink/white fls; ordinary well-drained soil, sun, invasive	0.08–0.3	P B
Hypericum olympicum		Jun–Aug	yellow fls; fertile well-drained soil, sun	0.23–0.3	P B
Lithospermum			funnel-shaped blue fls; sandy soil with peat/leaf mould, no lime, full sun	0.10–0.4	N
Lobularia maritima Alyssum maritimum	sweet alyssum	Jun–Sep	white/purple/lilac fls; ordinary well-drained soil, full sun	0.08–0.15	P
Saxifraga	saxifrage		many spp; most need sharply drained soil with lime and grit		N P
Sedum	stonecrop		many spp; numerous small star-shaped pink/yellow/white fls; well-drained soil, full sun	0.03–0.15	N P B
S. spurium		Jul	mat-forming plant, rich pink fls; well-drained soil, full sun	0.10	N
Thymus	thyme	Jun	several spp, small aromatic lvs variegated cvs, white/pink/lilac fls; well-drained soil, sun	0.03–0.23	N B
Veronica teucrium		Jun–Aug	sky-blue fls; well-drained soil, sun	0.23–0.5	N

CULINARY AND AROMATIC HERBS

Botanical name	Common name	Flowers	Notes	Height Metres	Value
Agastache foeniculum = A. anethiodora	anise hyssop	Jun-Sep	aromatic lvs used in cooking/pot-pourri; blue fls; ordinary well-drained soil, sunny posn	0.5	N
Allium schoenoprasum	chives	Jun-Jul	lvs used in cookery, pink fls; loamy soil, sun/semi-shade	0.15-0.25	N P
Borago officinalis	borage	May-Aug	edible lvs and blue fls; well-drained soil, sun	0.3-0.6	N P B
Marrubium vulgare	white horehound	Jul-Sep	medicinal herb, downy stems and lvs, fls in white whorls; poor soil	0.3-0.6	N
Mentha	mint	Jul-Sep	many spp, aromatic perennial used in cooking/pot pourri; best in rich moist soil, shelter and part shade	0.3-0.9	N
Ocimum basilicum	basil	Aug	aromatic lvs, used in cooking, small white-fls; warm posn	0.6-0.9	N
Origanum marjorana	knotted marjoram, sweet marjoram	Jul-Sep	aromatic lvs, used in cooking, white/mauve/pink fls; grow as an annual in well-drained soil, sunny posn	0.3	N
O. onites	perennial marjoram, Pot marjoram	Jul-Sep	aromatic lvs, pink/white fls; warm well-drained soil, shelter	0.6	N

Garden Plants Valuable to Bees

Botanical name	Common name	Flowers	Notes	Height Metres	Value
O. vulgare	wild marjoram	Jul-Sep	aromatic lvs, pink fls; well-drained calcareous soil, sun	0.3-0.6	N P B
Salvia officinalis	sage	Jun-Jul	aromatic lvs, used in cooking, purple/white fls; light well-drained soil, sun, shelter	0.6	N
Satureia	savory	Jul-Oct	aromatic lvs, used in cooking, rose/white fls; well-drained soil, sunny posn	0.3	N B
Thymus vulgaris	thyme	Jun	aromatic lvs, used in cooking, mauve fls; well-drained soil, sun	0.23-0.3	N B

BULBS, CORMS AND TUBERS

Botanical name	Common name	Flowers	Notes	Height Value Metres
Allium		Mar-Jul	onion-scented lvs, fls mauve/crimson/red/ yellow globular heads; well-drained soil, sun	0.15-1.2 N P
Anemone blanda		Feb-Apr	blue/mauve/pink/white fls; well-drained soil, sun/pt shade	0.15 P B
Camassia	quamash	Jun-Jul	racemes of star-shaped white/blue/purple fls; heavy moist soil, dead-head fls, plant 7-10cms deep	0.3-1.2 N P
Chionodoxa luciliae	glory of the snow	Feb-Apr	light-blue white centred fls; well-drained soil, full sun, plant 7-10cms deep	0.15 N P
Colchicum autumnale	meadow saffron	Aug-Sep	lilac/white/rose-pink fls; good in shrub borders/rough grass, plant 7-10cms deep	0.15 N P B
Convallaria majalis	lily of the valley	May-Jun	rhizomes, scented white fls; add leaf mould/compost to soil, pt shade	0.15-0.20 P
Crocus		Feb-Mar	many spp, fls yellow/white/mauve/purple; well-drained soil, shelter from wind, plant 7cms deep	0.05-0.13 P
C. balansae		Mar	orange fls	0.08 N P

Garden Plants Valuable to Bees

Botanical name	Common name	Flowers	Notes	Height Metres	Value
C. chrysanthus		Feb-Mar	golden-yellow fls, also cvs and hybrids in other colours	0.08	N P
C. tomasinianus		Feb-Mar	lilac/mauve fls	0.08	N P
Dahlia (single cvs)		Jul to frost	tuberous roots, bright coloured fls; well-drained soil with peat/compost, store tubers in frost-free posn	0.5-0.8	N P B
Eranthis hyemalis	winter aconite	Jan-Feb	tuberous roots, lemon-yellow fls; moist well-drained loam, sun/pt shade, plant 2cms deep	0.10	N P B
Fritillaria imperialis	crown imperial	Apr	pendent clusters of yellow/red/orange fls;fertile well-drained soil, sun/pt shade, plant 20cms deep	0.6-0.9	N B
F. meleagris	snake's head fritillary	Apr-May	chequered white/pink/purple bell-shapedfls; moist soil, good in short turf, plant 10-15cms	0..3-0.5	N P
Galanthus nivalis	common snowdrop	Feb-Mar	white fls with green markings; heavy moist loam, some shade	0.08-0.2	N P
Galtonia candicans	Cape hyacinth, Summer hyacinth	Jul-Sep	bell-shaped white fls; plant 20cms deep, leave undisturbed once established	1.2	N
Hyacinthella azurea = Hyacinthus azureus = Muscari azureum		Mar	spikes of pale blue fls; well-drained soil, pt shade good on rock garden, plant 5-7cms deep	0.08-0.2	P

Botanical name	Common name	Flowers	Notes	Height Metres	Value
Hyacinthus orientalis	hyacinth	Apr-May	hybrids in many shades, spikes of fragrant fls; plant bulbs 12-15cms deep	0.15-0.23	N P
Leucojum vernum	spring snowflake	Mar-Apr	white, green-tipped fls;: moist conditions, some shade, plant bulbs 7-10cms deep	0.20	N P
Muscari botryoides	grape hyacinth	Mar-Apr	spikes of blue/white fls; ordinary well-drained soil, full sun, plant 7cms deep	0.15-0.20	N P
Scilla sibirica		Mar-Apr	blue/white bell-shaped fls; moist well-drained soil, sun/ pt shade, plant 5-7cms deep	<0.15	N P
Tigridia pavonia	tiger flower	Jul-Sep	many cvs, fls usually yellow spotted crimson-brown; well-drained but moist soil, warm posn, sun, plant 7-10cms deep	0.5-0.6	N
Tulipa	tulip	Mar-May	fls goblet-shaped, many colours; alkaline soil, dead-head when petals fall, plant 15cms deep	0.15-0.9	P
T.kaufmanniana	water-lily tulip	Mar	greyish lvs, white fls flushed red and yellow, Many cvs of different colour; alkaline soil	0.10-0.25	P B

ANNUAL AND BIENNIAL PLANTS

Botanical name	Common name	Flowers	Notes	Height Metres	Value
Alcea rosea = Althaea rosea	hollyhock	Jul-Sep	tall spikes of fls, many colours; rich heavy soil, shelter, stake in exposed posns	1.4-2.7	N P B
Antirrhinum majus	snapdragon	Jul-Oct	spikes of fragrant fls, many colours; fertile light well-drained soil, sun	0.3-1.2	B
Begonia semperflorens	fibrous-rooted Begonia	Jun-Sep	red/pink/white fls; light moist but well-drained soil, pt shade	0.15-0.3	P
Calendula officinalis (single cvs)	pot marigold	May-Oct	orange/yellow daisy-like fls, lvs and stems with pungent aroma; poo/well-drained soil, dead-head regularly	0.6	N P B
Campanula medium	Canterbury bell	May-Jul	white/blue/pink/violet bell-shaped fls; fertile well-drained soil, sun/pt shade	0.3-0.9	N P
Centaurea cyanus	cornflower	Jun-Sep	sprays of pink/red/purple/blue/white fls; fertile well-drained soil, sun	0.3-0.9	N P
Cheiranthus	wallflower	May-Jun	freely produced fls, many shades; ordinary well-drained soil, sun, pinch out tips of young plants to encourage branching	0.15-0.6	N P B

Botanical name	Common name	Flowers	Notes	Height Metres	Value
Chrysanthemum Parthenium = Tanacetum parthenium	feverfew	Jul-Sep	short-lived perennial, aromatic lvs, white/yellow fls; fertile light well-drained soil, sun	0.23-0.5	B
Clarkia elegans (ingle cvs)	clarkia	Jul-Sep	spikes of white/lavender/pink/rose fls; medium-light slightly acid loam, sun	0.5-0.6	N P
Cleome spinosa	spider flower	Jul	white/pink/yellow fls; fertile well-drained soil, full-sun	0.9-1.2	N
Convolvulus tricolor =C. minor		Jul-Sep	blue/cinnamon/crimson fls; ordinary well-drained soil, sun, dead-head regularly	0.3-0.5	N P B
Cosmos bipinnatus	cosmos	Aug-Sep	white/pink/rose daisy fls; free flowering; light poor soil, sun	<0.9	N
Cucurbita pepo	ornamental gourds	Jul-Sep	large yellow fls, ornamental frs; ordinary, well-drained soil, train up trellis/trailing		N P B
Dianthus barbatus	sweet William	Jun-Jul	scented fls, many colours; ordinary well-drained soil, sun	0.3-06	N
Echium plantagineum =E. lycopsis	purple viper's bugloss	Jun-Aug	blue/purple fls; light dry soil, open posn, sun	0.9	N P
Eschscholzia californica	Californian poppy	Jun-Oct	orange-yellow, saucer-shaped fls; poor sandy soils, sun	0.3-0.5	N P B

Garden Plants Valuable to Bees

Botanical name	Common name	Flowers	Notes	Height Metres	Value
Gaillardia		Jul-Oct	bright yellow and orange fls; light, well-drained soil	0.5-0.8	N P
Gilia capitata		Jun-Sep	lavender-blue pincushion fl head; light well-drained soil, full sun	0.5	N P
Godetia (single cvs)		Jun-Aug	spikes/clusters of funnel-shaped fls; light moist soil, sun	0.3-0.6	N P
Gypsophila elegans		May-Sep	numerous small white/pink/rose fls; well-drained soil, sun	0.6	N
Helianthus annus	annual sunflower	Jul-Aug	huge yellow/copper bronze fls; well-drained soil, sun	0.9-3.0	N P B
Heliotropium X hybridum		Aug	small violet/lavender/white fls; fertile well-drained soil, sun	0.3-0.6	N B
Iberis umbellate	annual candytuft	Jul	small white/pink/purple fl heads; well-drained soil, sun	0.15-0.4	N P
Impatiens glandulifera	Himalayan balsam	Aug-Sep	purple/yellow/rose fls; fertile well-drained moist soil, sun/pt shade	0.9-1.5	N P
Lavatera trimestris = L.rosea	mallow	Jul-Sep	rose-pink fls; ordinary soil, shelter, sun	0.6-0.9	N P

Botanical name	Common name	Flowers	Notes	Height Metres	Value
Limnanthes douglasii	poached-egg flower	May-Jun	yellow and white scented fls; open sunny posn and cool root run – by paths/between rocks/ paving	0.15	N P B
Limonium sinuatum = Statice sinuate	statice	Jul	fls dried for decoration, many shades; well-drained soil, open posn, sun	0.5	N
Linum	flax	Jun-Aug	rose/blue fls; well-drained soil, sun	0.5-0.6	N P
Lobelia erinus	lobelia	May on	blue/white/pink fls; rich moist soil, pt shade, also trailing cvs	0.100.23	N
Lunaria annua = L.biennis	honesty	Apr-May	fragrant purple fls, silvery seed pods for indoor decoration; light soil, pt shade	0.8	N P
Myosotis	forget-me-not	Apr-May	blue small fragrant fls, loamy soil, pt shade	0.08-0.2	N P B
Nemophila menziesii =N. insignis	baby blue eyes	Jun	white centre sky-blue fls; moist soil, sun/pt shade	0.15-0.23	N
Nicotiana	tobacco plant	Jun-Sep	scented fls, many shades; rich well-drained soil, sun	0.6-0.9	N P B
Nigella damascena	love-in-a-mist	Jun-Aug	attractive seed pods, blue/white fls; fertile soil, sun	0.6	N P

Garden Plants Valuable to Bees

Botanical name	Common name	Flowers	Notes	Height Metres	Value
Oenothera biennis	common evening-primrose	Jun-Oct	large yellow fls; ordinary well-drained soil, sun	0.9-1.2	P
Papaver	poppy	Jun-Aug	fls in many shades; well-drained soil, sun		
Phacelia campanularia		Jun-Sep	lvs fragrant when crushed, blue fls; sandy well-drained soil, sun	0.23	N P
P. tanacetifoli	tansy phacelia	Jul-Aug	lavender fls; fertile moist soil, sun	0.6	N P
Reseda odorata	migonette	Jun-Oct	scented yellowish heads of small fls; rich well-drained alkaline soil, sun	0.5-0.6	N P B
Salvia hormium		Jun-Sep	coloured bracts, fls can be dried; well-drained soil, sun	0.5	B
Ursinia anethoides		Jun-Sep	colourful orange daisy-like fls; sandy soil, sun	0.5	P
Zinnia elegans (single cvs)		Jul-Sep	daisy-like fls, many colours, suitable for cutting/bedding; rich well-drained soil, shelter, sun	0.6-0.8	N P

26

PLANTS FOR THE WILD GARDEN

Botanical name	Common name	Flowers	Notes	Height Metres	Value
Agrostemma githago	corn cockle	Jun-Sep	purple fls; rare annual weed, once common in cornfields	0.3-0.9	P
Allium ursinum	ramsons	Apr-Jun	plant smells strongly of onion, white fls; damp, shady areas	0.3	N
Anchusa officinalis	alkanet	Jun-Aug	branched hairy perennial, purple fls; fields/ waysides etc	0.3-0.6	N
Anemone nemorosa	wood anemone	Mar-May	nodding, solitary white fls with pink of purplish tinge; woods/hedgerows	0.08-0.15	P B
Arctium	burdock	Jul-Sep	downy biennial, purple thistle-like fls; hooked Fruits, waste ground/dry woods	0.8-1.2	N P B
Asparagus officinalis	asparagus	Jun-Aug	fern-like lvs, tiny greenish fls; well-drained soil with compost	1.2-1.5	N P B
Arbarea vulgaris	winter cress	May-Jun	mustard-like perennial, dark shining lvs, yellow fls; well-drained soil with compost	0.3-0.6	N P B
Brassica			many spp esp crop plants if left to seed, e.g. cabbage, turnip, rape, mustard, etc		N P

27

Garden Plants Valuable to Bees

Botanical name	Common name	Flowers	Notes	Height Metres	Value
Campanula glomerata	clustered bellflower	May-Sep	stiff, hairy, perennial, clusters of violet fls; dry chalky fields	0.15-0.2	N P
Centaurea	knapweed Hardheads	Jul-Sep	purple thistle-like fls; rough grass/banks	0.3-0.6	N P B
Chrysanthemum vulgare = Tanacetum vulgare	tansy	Jul-Sep	attractive aromatic lvs, yellow button-like fls; fields/roadsides, very invasive	0.3-0.9	N P B
Clematis vitalba	old man's beard traveller's joy	Jul	rope-like stems, climber, cream fls; calcareous soil, hedges/woods		N P B
Cynoglossum officinale	hounds tongue	Jun-Aug	downy grey perennial, purple/red fls, hooked fruits; dry soil/sand-dunes	0.3-0.6	N P B
Daucus carota	wild carrot	Jun on	rough hairy biennial, feathery lvs, fls in whitish umbels; grassy places, alkaline soil	0.3-0.6	N P
Dipsacus fullonum	teasel	Jul-Sep	prickly biennial, conical lilac fl head with spines, standing through winter on dead stems; rough grass	1.8	N B
Echium vulgare	viper's bugloss	Jun-Sep	rough hairy biennial, fls vivid blue with pink buds; chalky/light dry soils, often near sea	0.3-0.9	N P B

28

Lightning Source UK Ltd.
Milton Keynes UK
UKHW051021071120
372987UK00007B/180